# TREES!
## HOW DO THEY GROW?

Olivia Watson
Illustrated by Tjarda Borsboom

Copyright © 2024 Hungry Tomato Ltd

First published in 2024 by Hungry Tomato Ltd
F15, Old Bakery Studios, Blewetts Wharf, Malpas Road, Truro, Cornwall, TR1 1QH, UK.

No part of this publication may be reproduced, stored in a retrieval system, or transmitted in any form or by any means, electronic, mechanical, photocopying, recording, or otherwise, without prior written permission of the copyright owner.

A CIP catalogue record for this book is available from the British Library.

ISBN 9781835693476

Printed in China

Discover more at
www.hungrytomato.com

**Words in BOLD can be found in the glossary.**

# Contents

| | | | |
|---|---|---|---|
| What Is a Tree? | 4 | Tree Protection | 16 |
| What Do Trees Need? | 6 | Brilliant Bugs | 18 |
| How Do Trees Grow? | 8 | Did You Know? | 20 |
| The Water Cycle | 10 | Match Up the Pairs | 22 |
| Spreading Seeds | 12 | Glossary | 24 |
| Types of Trees | 14 | | |

## Picture Credits

Abbreviations: m-middle, t-top, l-left, r-right, bg-background.

Shutterstock: Danny Ye 20bl; Digital Images Studio 19tr; Iakov Kalinin 23mr; Goinyk Production 23tl; Henri Kohkinen 18tl; Jeff Dalton 23tr; Kinggm.saleh 23bl; LizCoughlan 23br; malaka 18mr; nnattali 20tr; Olga Ilinich 21m; olko1975 19ml; Shootz photography 23ml.

Every effort has been made to trace the copyright holders, and we apologise in advance for any unintentional omissions. We would be pleased to insert the appropriate acknowledgements in any subsequent edition of this publication.

# What Is a Tree?

Trees are living things that can be found almost everywhere on Earth! They are important for people and nature. Trees come in all shapes and sizes, but most have the same four parts.

**Branches**
A tree's branches grow outwards from its trunk. It's here that leaves, fruit, flowers and nuts grow.

**Leaves**
Leaves are very important. They help the tree make its own food to give it energy and help it grow.

**Trunk**
The trunk acts as a drinking straw, carrying water and **nutrients** from the roots to different parts of the tree.

**Roots**
Roots are usually hidden underground. They help to hold the tree in place, like an anchor.

Apple seed

Twig

Fruit

Avocado seed

Some trees have other features, such as fruit, which are full of hidden seeds. Some are tiny, like apple seeds, and some are much bigger, like avocado seeds.

## Not a tree

Some plants that we call trees aren't trees at all! Cacti and boojum aren't made of wood, and palms don't have branches, so they don't count!

Cactus

Boojum tree

Palm tree

# What Do Trees Need?

**Trees can't grow without a few very important things: sunlight, air, water, and nutrients.**

### Sun
Trees take light from the Sun and turn it into food, which gives them energy to grow.

### Water
Without enough water, trees would shrivel up and die.

### Air
Just like us, trees need plenty of air to stay alive. They use it to make food!

### Nutrients
A tree's roots take up nutrients and water from the soil that it needs to grow.

### Insects
Bugs like bees and butterflies **pollinate** flowers, which helps trees grow fruit and seeds.

### Space
Trees need lots of space around them to avoid **competing** with other trees for nutrients and sunlight.

### Wildlife
Wildlife and trees help each other. Animals help trees spread their seeds, while branches and trunks make great woody homes.

# How Do Trees Grow?

**Every tree starts its life as a tiny seed, even the most gigantic ones! How does it all begin?**

## Seeds

Seeds come in all shapes and sizes, but they're all packed with food to keep the **seedling** alive until it grows and can make its own.

**1.**
Once planted and given water, a root starts to grow out from the seed. This is called **germination**.

**2.**
A shoot grows up from the seed towards the soil's surface as the root grows down.

**3.**
The roots keep growing as the shoot becomes a trunk. A young tree is called a sapling.

**4.** The tree grows bigger and stronger, and starts growing flowers and fruit on its branches.

**5.** Eventually, the tree gets weaker and stops growing. Even **decaying** trees make great homes for wildlife.

# The Water Cycle

**Trees need water to stay healthy and strong, but they also help water to be recycled across the world. We call this the water cycle.**

### 2. Leaves
Even rainwater sat on leaves is swept up. Trees release extra water through their leaves if they have too much, too!

### 1. Evaporation
The Sun warms water from oceans and rivers, turning it into tiny droplets that rise into the air.

## 3. Condensation
Up in the sky, the water droplets cool down and stick together as clouds.

## 4. Precipitation
When clouds get very heavy, the water droplets fall back down as rain, snow, or hail.

## 5. Collection
Plants and rivers collect water that runs into the soil or ocean, then the cycle starts again!

# Spreading Seeds

**Trees can't move around like animals, so they have to find clever ways to spread their seeds.**

Willow tree

## Up, up and away
Some seeds are swept up and scattered by the wind. Winged seeds can fly far!

Silver birch tree

## Water surfers
Some trees, like willows, lean over streams and rivers. Their seeds fall into the water and are carried away by the **current**.

## Seedy surprise
Some animals eat the fruit from trees. They poop out the seeds they can't **digest**, leaving them to grow into new trees.

## Exploding seeds
Some trees have **seedpods** that explode! Bursting open, they shoot their seeds as far away as they can.

Sandbox tree

## Sticky seeds
Seeds with tiny spikes or hooks stick to animals' feathers or fur. They can be carried a long way before falling off.

## Hidden treasure
Squirrels bury nuts and seeds, saving them for winter. The ones they forget about grow into trees!

# Types of Trees

There are more than 60,000 types of trees in the world! They all fit into a main group based on how they grow: deciduous or evergreen.

### Deciduous
Deciduous trees lose their leaves in autumn and regrow them in spring.

### Broadleaf
Most deciduous trees have wide, flat leaves.

Ash tree

Magnolia tree

### Flowers and fruit
Only deciduous trees grow flowers! They can be all shapes and sizes, but only bloom for a short time every year.

### Hardwood
Hardwood is another name for deciduous trees. They usually have heavy trunks and grow much slower than softwoods.

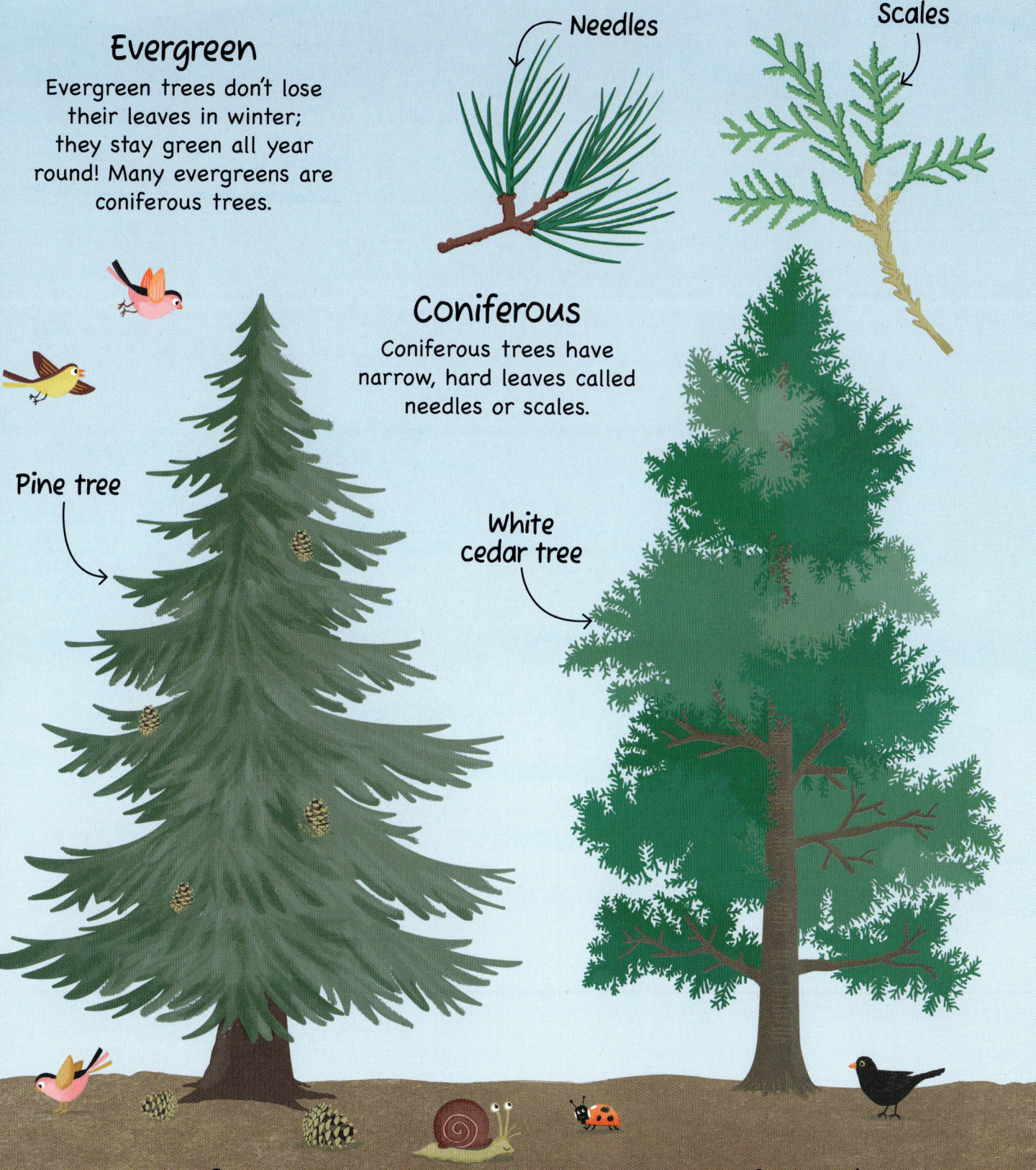

# Tree Protection

Trees have clever ways of protecting themselves from the animals that try to hurt them, but sometimes they need a helping hand.

### Acacia alert
When acacias are being eaten, they send out a warning! Nearby acacias release bitter **tannins** into their leaves to stop them being eaten next!

### Helpful elephants
These gentle giants love to munch on young trees. This stops too many trees from competing for nutrients and space, meaning older trees can keep growing taller.

### Undercover chemicals
Trees like oak and spruce also release tannins to make insects that eat them ill. What a nasty surprise!

### Sap trap
Some trees, like lodgepole pines, release sticky **sap** that traps bugs, stopping them from eating the **bark** and killing the tree.

### Food chain
The food chain keeps nature balanced; it stops **herbivores** from eating too many plants and trees.

# Brilliant Bugs

Did you spot the creepy-crawlies hidden throughout this book? Creepy-crawlies play an important part in keeping trees healthy.

## Predatory protectors

Some insects, like lacewings, are **predatory**; they hunt tree-damaging insects, like caterpillars and mites. By doing this, they protect trees from **infestation** and disease.

## Perfect pollinators

Many trees rely on flying insects like bees, butterflies, and beetles to pollinate them, which helps the tree create seeds and grow more new trees!

Bees are some of the best pollinators!

## Sickness killers

Certain beetles and moths only eat stressed or already dying trees. By eating these trees, the bugs speed up **decomposition**, make space for new trees to grow and stop the disease from spreading.

Jewel beetles help with decomposition.

Weevils are great for the environment!

## Ecosystem balancers

Sometimes plant-eating insects are good! By munching on **invasive** plants, these insects stop the new plants from taking over, which keeps forests balanced.

## Nutrient recyclers

Insects like termites, wood-boring beetles and worms keep soil healthy by breaking down dead plants and animals, recycling the nutrients into the soil. This helps trees get the nutrients they need to grow.

# Did You Know?

Trees are pretty amazing! Every living creature needs trees to survive; the world wouldn't be the way it is today if we didn't have them. Did you know these amazing facts about trees?

A large oak tree can drop as many as **10,000 ACORNS** in one year!

Magnolia are one of the oldest known **FLOWERING PLANTS.** In the times before bees existed, flowers were pollinated by prehistoric beetles instead.

Bamboo isn't actually a tree - it's a **GIANT GRASS!**

The Coulter pine grows some of the
# BIGGEST CONES
of all. These giants can weigh as much as 5 kg (11 lbs)!

The Empress tree is one of the
# FASTEST-GROWING
trees in the world! It can reach 6 m (20 feet) in its first year.

Every year, people from all over the world travel to Japan for Hanami, the ancient tradition of enjoying the
# BLOOMING OF CHERRY BLOSSOMS.

# Match Up the Pairs

**Can you match up the fact boxes (below) with the correct tree (right)? Flip back through the book if you need a hint!**

### 1.
I have "tree" in my name, but I'm not actually a tree!

### 2.
I'm a coniferous tree that grows cones, not delicate flowers.

### 3.
I'm covered in spikes and have exploding seedpods.

### 4.
I can make my leaves bitter to stop giraffes from eating me!

### 5.
My fluffy-looking seeds fall into rivers to be spread by the current.

### 6.
I'm a hardwood tree with big, beautiful flowers.

Pine tree

Willow tree

Magnolia tree

Palm tree

Sandbox tree

Acacia tree

Have you matched them all?
Answers can be found on page 24.

# Glossary

**Bark** – the tough outer layer of a woody plant stem or root, such as a tree trunk.

**Competing** – (verb) going against one another to gain or win something.

**Current** – the continuous movement of a body of water, such as a river or ocean.

**Decaying** – (verb) slowly breaking down over time. This happens to plants and animals after they've died.

**Decomposition** – the process of a living thing breaking down after it's died.

**Digest** – (verb) to break down food into substances that can be absorbed and used by a body or plant.

**Germination** – the process when a seed begins to sprout roots and shoots.

**Herbivores** – animals that only eat plants.

**Infestation** – when a large number of animals live where they're not wanted.

**Invasive** – something that takes over a place it's not meant to be and causes harm.

**Nutrients** – substances or ingredients that plants and animals need to live and grow.

**Pollinate** – (verb) the process of moving pollen (see below) from one flower to another – often by an insect – so the plant can make new seeds.

**Pollen** – a dusty power made by some plants. It is used to produce new seeds.

**Sap** – a watery substance that comes out of a plant or tree.

**Seedling** – a young plant, grown from a seed.

**Seedpods** – pouches or cases produced by some plants to hold their seeds.

**Tannins** – bitter-tasting chemicals naturally found in many plant leaves.

## Answers to Match Up the Pairs

Answers: 1. Palm tree, 2. Pine tree, 3. Sandbox tree, 4. Acacia tree, 5. Willow tree, 6. Magnolia tree.